Out of This World: The Movement Dimension

By

Timothy Michaels

Table of Contents

Preface

As an artist and student of science, I am very interested in learning how the world works.

Studying physics, I find information that scientific research has revealed, and the interpretations of that information. I try to separate the two, and understand what is known, and what is derived from that information. Perhaps this is because much of it seems so strange that I want to know how reliable it is.

Many people are attracted to science in search of strangeness or fantastic and unbelievable possibilities. My interest is in understanding my world, not regarding the possibilities of technology, but nature itself. I want to know how principles of nature affect my art and my life in general, and what they say about the world around me.

In searching for how the world works, I've seen some of our most popular and well-established theories point toward what I believe is the foundation of science. But, the interpretations of those theories as they currently are, seem incorrect. There has been difficulty in assembling relativity and quantum theories, which I believe is related to their interpretations.

I think I've found a way to assemble these pieces which makes sense. It helps me to understand nature, and could also lead to even deeper knowledge of the universe. This is what I intend to share in this book.

For those readers who aren't familiar with some of the progress that we've made in science, I'll review it so they can understand the information that I'm assembling. Then I'll propose a solution which I believe is a good fit and answers more questions than previous interpretations have.

If I'm correct, where this leaves us may be very close to the fundamentals of the universe. Though there will still be more data to gather and details to explain.

There will also be simplifications, as I believe the laws of our universe, at its most fundamental level, should not be as complicated as they currently appear.

Essentially what I want to do is point to ideas which are already well known and accepted. Then explain why they are fundamental, how we should look at them, and how we can proceed to understand their relationships.

What I'm offering the reader is an explanation of some of our greatest theories for those who aren't already familiar with them, the entertainment of how surprising they are, and a potential solution to what they mean, which is nothing short of amazing.

Chapter 1 - Introduction

Movement is about all events. Everything depends on it.

It's not just how we walk across a room or get to and from work. It's responsible for the orbiting of planets, the weather, digestion, the flow of electricity through a wire, the feeling of pain making its way to our brain, and thought.

It isn't the actual events, but rather what is behind the events. Change is what they are based on, and movement is the process of change. It is also what our laws of physics are about. They specify and describe movement.

The topic appears simple, but we've been learning about it for centuries and recently it seems to have become incomprehensible. Theories of movement no longer look like they describe the world we live in. They seem to go beyond the dimensions of space and time which we're familiar with. I believe this is for good reason. I believe that movement is literally "out of this world."

The difficulty comes from advancements made in physics about 100 years ago which we're still trying to make sense of. Specifically, relativity and quantum theories. If you're familiar with them, you know how "other-worldly" they seem to be. Yet their formulas are extremely accurate in predicting movement in our world. So accurate, that we've been able to make tremendous technological advancement by applying them.

In an interview with the American Association for the Advancement of Science, Richard Feynman, while defining science, said "...what happens in nature is true and is the judge of the validity of any theory about it."

With this guidance in mind, and looking at the scientific theories we have today, I want to know how they make sense with my daily experience. I want to see science judged by nature and be believable.

Much of the strangeness of these theories could be due to the way they're currently described. The difficulty with them may not be in our ability to understand them as much as it is in our ability to believe them.

My purpose here is to suggest how these great theories could make sense in our world. I'll be putting them together in a model which very much does look like the reality we're familiar with. And, while it too is unusual, I think it makes sense with some simple explanation. Hopefully the result is that many people come to understand our reality and that science can use this model to advance further.

First, I'll review relativity and identify its characteristics. Then, I'll review quantum theory and look at its qualities. I will point out how they seem to conflict and then show how these behaviors can all fit together in a realistic way with the world we know.

Chapter 2 - Relativity

Galileo Galilei (1564–1642) had a theory about relativity which stated that wherever we are (our frame of reference), as long as we're moving uniformly, the laws of motion will be the same. He believed that the natural state of motion is uniform motion, not stillness. This means that pouring a glass of wine on a steadily moving airplane works the same as pouring one in our dining room. Uniformly moving just means that whatever movement there is, it is so steady that we don't know it's occurring.

For example: The earth completes one rotation every day. That means that if you're at the equator, you're moving at 1,038 mph. The earth also orbits the sun at a speed of 66,900 mph. The solar system moves 500,000 mph through the galaxy. And the galaxy moves 1,386,000 mph through the universe. We don't feel any of this because it's smooth. For that reason, it doesn't affect the way things move around us.

In 1905, Albert Einstein updated the theory of relativity to specify that the laws of "physics" are the same for any reference frame in uniform motion. This is what we know today as his special theory of relativity. It wasn't dramatically different. He only changed laws of "motion" to laws of "physics". The statement remained essentially the same. But, he explained the consequences of it, and those are dramatic.

The Basic Concepts

The speed of light is constant, and time and distance adjust themselves to accommodate that.

Galileo, too, believed the speed of light was constant, but couldn't account for how that affects other objects. Einstein had the benefit of James Clerk Maxwell's explanation of how energy moves (this included light). In 1887, Heinrich Hertz had proven Maxwell to be correct.

So the time had come for relativity to describe the consequences of very fast motion.

Maxwell's explanation of movement of energy was that energy, such as light, moves itself. It's not propelled like many things we're familiar with.

If I throw a ball, its motion is relative to my arm. And, other things are related to the ball by their relation to my arm. Energy, however, throws itself. It propels itself. It does this with its electric and magnetic fields. The electric field creates a magnetic field. And that magnetic field creates an electric field. Then the electric field creates another magnetic field. And this pattern repeats.

The movement of energy is not based on where it originated. Nothing threw it. So its movement is related to everything we want to compare it to as it "crawls" through space, very quickly.

You can observe how a changing electric current creates a magnetic field. Just hold a magnetic navigational compass near a wire carrying electricity and you will see it respond, as Hans Christian Oersted did in 1820.

And you may observe a magnetic field creating an electric field by waving a magnet past a wire and measuring the very tiny electric current it causes in the wire, as Michael Faraday did in 1831. This is how an electric generator works, a dynamo, like those Einstein's father was in the business of building.

The way Heinrich Hertz proved the propagation of energy was by creating a spark on one side of a room and positioning a circular wire on the other side of the room to receive the energy from it. The coiled wire had a gap in it and was not connected to anything. A spark created on one side of the room caused a spark in the gap of the unassociated wire coil on the other side. He had transmitted energy through the air.

4

Nicola Tesla applied this principle when he built the first patented radio. Then Guglielmo Marconi used it in 1901 when he sent signals from England to North America. Today it's applied to radio, television, cell phone, and other wireless devices for communications.

Maxwell realized that this electro-magnetic energy he had been studying was the same as light when he calculated its speed and found it was a match.

The speed of light was known very precisely by then (within 0.05% of its exact value), thanks to Albert Michelson's measurement in 1879 with one of his interferometers (devices which detect wave interference or misalignment).

With the benefit of this research done by his predecessors, Einstein was able to say that since the speed of light is predicted (could be calculated) by the laws of physics, then the speed of light should be the same in all reference frames according to relativity. And, he accepted the consequences of this.

It is significant that Maxwell was able to calculate the speed of light without the need to measure it. This is how we know that it's constant. The speed of light is fixed as a result of its propagation, regardless of conditions.

The consequences of light speed being fixed are that the speed of light measured by a person moving toward the light source is the same as the speed of light measured by a person moving away from it. This is true even if he was moving very fast, like near the speed of light himself.

Light doesn't care where it came from. It moves relative to us regardless of how we're moving. We might imagine that if we're on a moving plane it gets on the plane with us.

If I was in a car moving at 20 mph, and I threw a ball forward at 30 mph, I would see the ball moving forward at 30 mph and I would not notice the car moving at all (because I'm in it).

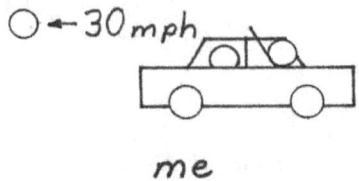

me

A person on the side of the road would see the car moving forward at 20 mph and see the ball moving forward at 50 mph (speed of car plus speed of throw).

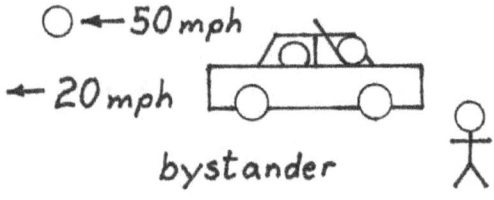

bystander

If I threw a ball backward at 30 mph from a car which is moving forward at 20 mph, I would see the ball going 30 mph backward and I would not notice the car moving at all.

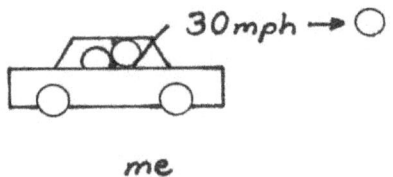

me

A person on the side of the road would see the car moving forward at 20 mph and see the ball moving backward at 10 mph (speed of car minus speed of throw).

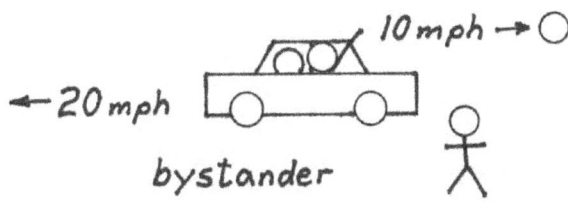

This is because the speed of the ball is always based on the thrower of the ball. But, we don't throw light. It moves itself in relation to us.

I could be riding on a rocket with a flashlight, traveling almost as fast as light in any direction, and the light coming from my flashlight would still be the same speed. It would be 670,616,629 mph for me, for a bystander, and for anyone who cares to measure it.

In order to make this behavior of light fit into our world, to make sense of it, we say that time and space adjust themselves. They bend or warp. That's the popular explanation for the constancy of light and varied experience of time and distance.

Space and Time

In 1908, Einstein's former professor, Hermann Minkowski (1864-1909) suggested that time is a dimension and that space-time is how we should describe our world.

Here's how space and time accommodate a constant speed of light. For simplicity, I'll round the speed of light to 700 million mph.

If I travel a distance of 700 million miles at half the speed of light (350 million mph) while shining my flashlight ahead of me, a person standing on the side of the galaxy sees the light get to its destination (700 million miles away) at the speed of light (700 million mph).

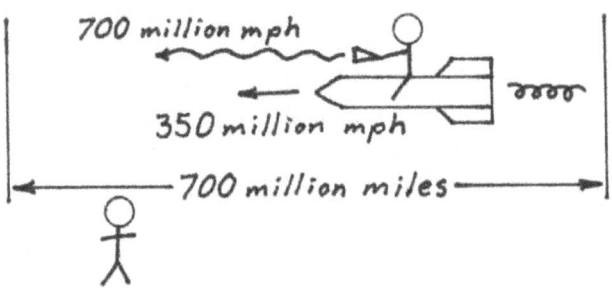

So, the bystander sees the light arrive in 1 hour. The bystander sees me, traveling at half the speed of light, arrive in 2 hours.

But, from my point of view (riding on the rocket), the light is moving faster than me by the speed of light. And, I'm moving at half the speed of light. (Really, the destination appears to be coming to me at half the speed of light.)

So according to my watch, I get there in 1.73 hours and, according to my odometer, I've only gone 606 million miles to get there. I see the light arrive there in 0.87 hours, half the time as me, because it's going twice as fast.

My experience of time and distance while riding on the rocket, is less than that of the bystander by about 13%.

From the light's point of view, it gets there in zero time and has traveled zero distance. It experiences the world as being in all places at all times.

The mathematical formulas are as follows:

$$T_1 = T_2 \cdot \sqrt{1 - (travelspeed \text{ as } \% \text{ of lightspeed})^2}$$

T_1 = Time experienced by traveler
T_2 = Time observed by bystander

$$D_1 = D_2 \cdot \sqrt{1 - (travelspeed \text{ as } \% \text{ of lightspeed})^2}$$

D_1 = Distance experienced by traveler
D_2 = Distance observed by bystander

In the previous example:

Since the time (T_2) that the bystander sees me travel is 2 hours, and my speed as a percent of light speed is 0.5, then my experience of time is 1.73 hours.

$$2 \text{ hours} \cdot \sqrt{1 - 0.5^2} = 1.73 \text{ hours}$$

The distance (D_2) that the bystander sees me travel is 700 million miles, and my speed as a percent of light speed is 0.5, so my experience of distance is 606 million miles.

$$700 \text{ million miles} \cdot \sqrt{1 - 0.5^2} = 606 \text{ million miles}$$

Since the time (T_2) that a bystander sees light travel is 1 hour, and the speed of light is 100% of light speed (or 1.0), then the light's experience of time is zero hours.

$$1 \cdot \sqrt{1 - 1^2} = 0 \text{ hours}$$

The distance (D_2) that a bystander sees light travel is 700 million miles, and the speed of light is 100% of light speed (or 1.0), so the light's experience of distance is zero miles.

$$700 \text{ million miles} \cdot \sqrt{1 - 1^2} = 0 \text{ miles}$$

The reason we didn't notice how movement affects our experience of time and distance long ago is because it's extremely minimal at our slow speeds.

This "dilation", or shortening, of time and distance applies to both energy and matter as they move, but is only noticeable at high speeds.

EXPERIENCE OF TIME & DISTANCE

Object	Speed	Experience of Time & Distance
Person running	8 mph	near 100%
Driving car on highway	75 mph	near 100%
Airliner	500 mph	near 100%
Earth rotating	1,038 mph	near 100%
Earth orbiting sun	66,900 mph	near 100%
Lightning	270,000 mph	99.999992…%
Sun moving through galaxy	500,000 mph	99.999972…%
Galaxy moving through universe	1,386,000 mph	99.999786…%
Electron in ground state orbit	4,895,000 mph	99.9973…%
Some quasars	604,000,000 mph	43.6….%
Light	670,616,629 mph	0%

For those who enjoy speed, be careful not to go too fast otherwise you won't be able to experience it. Any more than a couple hundred million miles per hour just isn't worth it.

Mass & Energy

Based on relativity, Einstein found energy and matter (mass) to be equivalent. And the formula for converting from one to the other is as follows:

$$E = mc^2$$

E = Energy in joules (1 Calorie is 4190 joules)
m = Mass in kilograms (1 pound is 0.454 kg)
c = Speed of light (299,792.458 meters/second)

To find mass, the formula would be re-arranged as:

$$m = E/c^2$$

Matter, a real tangible thing, is equivalent to energy, which is invisible and seems entirely different from tangible objects. This may not be too hard to understand though, because we experience it all the time. Or we think we do.

A candle is tangible. It has weight. It is matter and has mass. But after burning it, it's mostly or entirely gone. What we got out of it was light and heat, which are energy.

A full tank of gas in your car is heavy, but after driving a long way the gas is gone. The weight (mass) is gone. What you got out of it was movement. It did a lot of work by converting to energy.

What's deceiving about these examples is that the entire mass of the candle and entire mass of the gasoline did not convert to energy. Only a little bit of it did, by way of chemical reaction (molecules turning into different molecules).

If you could gather all the smoke and gas (fumes) produced as the candle burned and your fuel burned, you would have collected nearly the same amount of mass as you started with. Only a tiny bit of it actually transitioned from matter into energy.

Radioactive decay, which is used in nuclear power plants, is much more efficient at converting matter into energy. The process leaves behind less waste material. This shows better the relationship between matter and energy. A small amount of matter is equivalent to a whole lot of energy. However, it too is only converting a small percentage of matter into energy. Some of the mass of nuclei (center parts of atoms) is converted, but the atoms all remain, and are just slightly smaller.

An example of 100% conversion of matter into energy is the power source of Star Trek's USS Enterprise. It combines matter with anti-matter, which results in them both being fully annihilated, leaving only energy. While particles and their anti-particles do annihilate each other all the time in real life, these are extremely small events which we are not yet able to control at any scale or efficiency to make it usable as a power source.

But, if we could access all of the energy contained in matter we would find that an 8 ounce steak, which is 0.2268 kg, multiplied by the speed of light (299,792.458 m/s) squared gives us 20,380,000,000,000,000 joules of energy, which is 4,865,000,000,000 Calories.

At 100% efficiency, an 8 ounce steak would yield almost 5 trillion Calories. That's enough to feed the population of the United States for two years.

If we could make full use of $E = mc^2$, we would only need to eat 2.7 mg to live a lifetime. That's a grain of sand.

However, our bodies are highly inefficient. We mostly convert matter chemically into other matter. We don't actually change every molecule into energy. That allows us to eat much more food.

Gravity

Gravity is not a force, but a warping of space-time.

In general relativity, not just constant motion but constant acceleration is included, and considered a uniformly moving reference frame. This is due to inertial mass (the heaviness felt during acceleration) being equivalent to gravitational mass (the heaviness felt from gravity).

In 1907, Einstein said that something accelerating is equivalent to something which is still, but is influenced by gravity. This contributed to his theory of general relativity in 1915, which explains that gravity is not a force but is a curvature of space and time.

In special relativity the experience of space and time shrink as speeds approach the speed of light. Now, we see shrinkage of space and time due to the equivalence of movement. Just the effect of acceleration, which gravity is, is causing the experience of space and time to be reduced.

Regarding gravity being a curvature of space-time, it's unproven. But, regarding Einstein's prediction of ripples in the structure of space-time being created by moving bodies, some progress has been made.

The Laser Interferometer Gravitational-Wave Observatory has interferometers set up in Washington and Louisiana. Each has 2 perpendicular 2.5 mile long arms capable of detecting length changes as small as 1 trillionth of the diameter of a human hair. In 2015, both detectors simultaneously twitched for 20 milliseconds. This has been calculated to be a potential gravity wave resulting from 2 black holes orbiting each other, then colliding at nearly the speed of light 1.3 billion years ago.

Another project has been proposed which would place stations in space forming a triangle 3 million miles on each side and be able to detect changes of 1 billionth of an inch.

And general relativity's predictions of the effects of gravity on both matter and energy are well proven.

As part of his presentation of general relativity to the Prussian Academy of Sciences in 1915, Einstein explained the changes in the orbit of the planet Mercury. These changes had been a mystery for centuries and had been unresolved by Newton's theory of gravity.

In 1919, during a solar eclipse, Arthur Eddington and others observed the bending of light around our sun in the form of stars appearing out of place as they passed behind the sun, yet were visible at its edge, where Einstein predicted. Light from distant stars bent toward the sun as it passed near it.

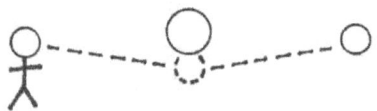

And, today, a cluster of 10,000 galaxies, which are 2 billion light years away and contain about 100 billion stars each and perhaps 100 billion planets each, is used as a massive gravitational lens.

It bends light from stars which are even further away, providing us with 30 times magnification and allowing us to see deeper into space. It's really no different from a magnifying glass that we're familiar with.

It too, is actually made of a large quantity of particles with lots of empty space between them. But, while the particles of a magnifying glass are held together by electric charge, the galaxies are held together by gravity.

Continuum's

A continuum is something which is infinitely divisible, and does not have a minimum measure. It's something that is continuous.

Emmy Noether (1882-1935) found mathematically, that for the rules of a logical system to be consistent, there must be a continuum. And where there is a continuum, the rules must be consistent.

While derived mathematically, I view this as being an aspect of relativity. One could have derived the continuum requirement from Einstein or Galileo's relativity. A continuum is a necessary part of a uniformly moving reference frame. Relativity requires smoothness. And laws being consistent, mean that they work the same, which is what relativity states.

In relativity, smoothness means a feeling of steady, or uniform motion. Smooth motion is only possible in a time and space which is not choppy, but continuous.

Noether's theorem, whether we consider it as relativity or not, stands on its own as a successful theory. It has contributed to finding relationships between fundamental particles and also to black hole research.

Chapter 3 - Quantum Theory

The consequences of relativity are strange and hard to imagine as being a part of our world, yet its predictions are very accurate.

Quantum theory is even more strange and more accurate. In fact, quantum theory is the most proven theory in all of history. With an accuracy rate of 99.9999998%, our world economy today depends on it. It calculates the workings of transistors, microchips, lasers, superconductors, microwave ovens, cell phones, televisions, radar, and the internet.

Quantum theory is a set of theories which came out of the discovery that our world is quantized. It began with the finding that energy is only available in certain size chunks, or packets, which can't be divided further.

This led to the realization that there is also a minimum measure of distance and a minimum measure of time. Everything in our world comes in building blocks so small we hadn't previously noticed them. This means that our world could be described as a very high-resolution digital environment.

Here are some of the most significant aspects of the quantum theories.

Quantization

The Merriam-Webster dictionary defines "quantized" as "to subdivide (something, such as energy) into small but measurable increments."

In 1900, Max Planck declared that energy, such as light, exists in the smallest quantities which he called "quanta" and cannot be divided further. This was the beginning of quantum theory.

In 1926, Gilbert Lewis named a quantum of light a "photon". As a result, light energy now goes by either name. In fact, all electromagnetic energy does, including X-rays, television, radio, and cosmic rays.

Building on this quantization, we eventually found distance and time to be quantized, too. Our dimensions of space and time are grid-like.

A quantum of light energy (photon) is about 10 million waves long.

The smallest measure of distance is called a "Planck", and is about 10^{-35} meters long (10^{-35} is a 1 with 35 zeros in front of it, 0.00000000000000000000000000000000001 meter). It's the distance light travels in one Planck (named for Max Planck) of time.

The smallest measure of time is also called a Planck, which is 0.0000000000000000000000000000000000000000001 seconds (10^{-43} seconds). This is the time it takes light to travel one Planck of distance.

This quantization has a lot to do with making our world the way it is. It prevents atoms from collapsing by keeping electrons away from the nucleus and in designated orbits. (Neils Bohr explained this in 1913.)

This is important because atoms are 99.9999999999999% empty space. If atoms collapsed, the earth would only be 419 feet in diameter. Even worse, the whole universe would be nothing but rocks with no interesting properties. The properties of our chemical elements are a result of the behavior of electrons in their orbits.

Wave Functions

At the subatomic scale, individual particles seem to behave randomly. So, in order to make predictions, which is a primary objective of science, we now assume all particles to exist as what we call a wave function, or a probability cloud.

This idea was first proposed by Max Born in 1926. He said that this wave aspect of a particle is a function of position, a wave function, which tells us the probability of finding a particle at a location. This may be imagined as a typical bell curve in which we're most likely to find a particle where the curve is highest. And the lowest parts of the curve extend infinitely far in all directions, never quite reaching zero.

Today, as part of quantum theory, we assume that a particle could be anywhere and it's not our job to find where it is, but where it most probably is. We imagine this as a cloud of possibilities which spans the entire universe, but which has a concentrated density where it's most likely to be.

We calculate all of these infinite possibilities and achieve great accuracy. It's this accepting of all possibilities which leads us to the most accurate predictions in science. Much of the mathematics for calculating these infinite possibilities were developed by Richard Feynman and Julian Schwinger, and also separately by Sin-Itiro Tomonaga. The formulas in popular use today as the foundation of modern electronics are by Feynman and Schwinger.

Particles & Waves

Everything is a particle and a wave.

Quantum theory describes the particle aspect of objects, and also the wave aspect of the same objects. It describes everything as both a particle and a wave. This has a lot of consequences.

Particles are real tangible objects which have a defined form and location. This is generally considered to be the true description of all things, that all things are actually particles.

The wave aspect of objects is defined in 2 ways:

1) All objects are said to have a location which is not entirely certain but can be explained as a range of probabilities (the bell curve previously described). This is not considered to be a real quality of objects, only a model which accurately predicts their location.

2) The second way objects are described to be wave-like is in their interactions. We know that waves which cross paths can enhance or diminish each other, affecting where they appear to be located during their interaction. This behavior shifts the probability of finding them. This too, is not considered to be a real description of objects, only a way of explaining their behavior.

What this means is that everything, whether it's matter or energy, liquid or solid, magnetism or light, can be located as a particle. As a result, it can impact objects as particles do. Everything can also do things which waves do, like interfere with each other and spread out across a distance.

The realization of this began with light. As far back as ancient Greece, philosophers argued over whether light is a particle or a wave. (We didn't really have scientists, then. The field was called "natural philosophy".)

- Democritus (460-370 BC) created the term "atom" and was considered an "atomist". This meant he believed there is a smallest particle of matter called an atom. He also believed that light is made of particles.

- Aristotle (384-322 BC) was not an atomist and believed light is due to waves in a medium, the ether, which permeates all of space. It was he who declared the ether to be the 5th element (earth, air, fire, water, ether).

- In 1678, Christian Huygens proposed that light is waves of different frequencies, like sound is.

- Isaac Newton, in his 1704 book, *"Optiks"*, described light as particles.

- In 1801, Thomas Young proved that light is a wave. He did this by shining light through 2 pinholes and observing an interference pattern created by the 2 beams of light on a screen. Interference is when waves either align or misalign. This results in them becoming stronger or weaker.

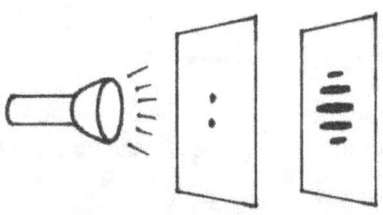

This may be illustrated in 2 ways:

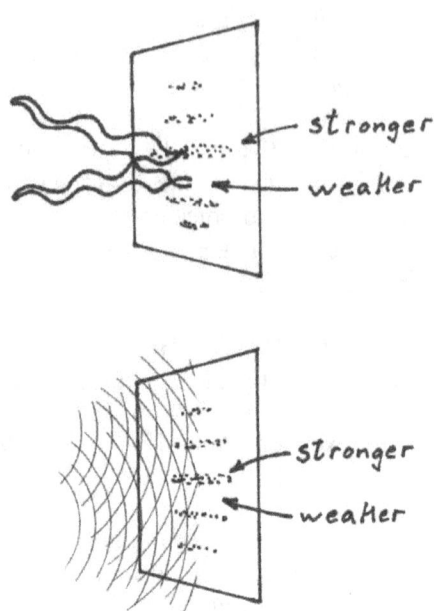

The visible result on a screen is bands of light where waves enhance each other (constructive interference) between bands of dark where waves cancel each other (destructive interference).

Taking advantage of the wave property of light, Albert Michelson had been using its interference to measure time. Knowing the speed of light and length of its waves, he could observe interference to reveal slight variations in time.

All the while, those who believed in the waves of light also believed in the ether as an invisible substance for the waves to travel through.

So, going further in proving that light is waves, Michelson and Edward Morley set out in 1887 to prove the existence of the ether. Instead, they found that it doesn't exist.

What they did was to separate a beam of light into 2 parts, then rejoin them at a screen to create an interference pattern. The way they used this to detect for the ether, was to have the 2 beams going different directions before rejoining.

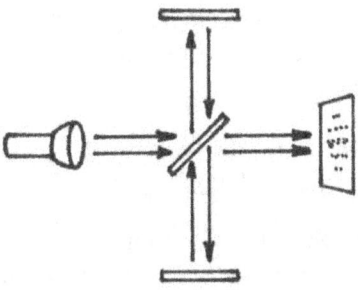

The light was directed at a half-silvered mirror which reflected half of the beam and allowed half of it to pass through. The light passing through it went straight to a screen. The reflected beam went off to the side to be reflected back at the mirror. What passed through it this time got reflected back again and eventually got directed at the screen.

This certainly would assure that the 2 beams went different directions and could be misaligned to produce an interference pattern. The key was the direction the light traveled to get to the screen. Some went straight. Some went straight and side-to-side.

An assumption was made. If there is an ether, we certainly would be moving through it, somewhat like walking through water. The waves around us should move faster or slower in different directions as we move.

Since the earth is rotating, and it's orbiting the sun, and the sun is moving around our galaxy, and our galaxy is moving through the universe, then certainly if there is something which is in all places which light travels through, we must be moving in relation to it.

The Michelson–Morley device was set up to very precisely detect changes in the speed of light going forward, compared to going side-to-side. Whatever the interference pattern was initially didn't matter. They just needed one they could watch.

Then, they slowly rotated the table and watched for a change in the pattern. This would have indicated a change in the relative speed between forward and side-to-side light. No change was detected. There was no ether.

With this experiment, the belief that light had a medium to travel through was gone, after 2300 years.

In the same year (1887), Heinrich Hertz observed the photoelectric effect in which light seemed to be a particle. He shined light at metal and saw electrons being knocked out of the metal, like BB's knocking chips of rock out of a wall.

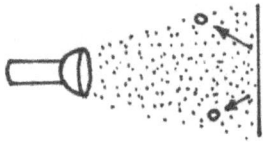

But, in 1905, Einstein wrote his disagreement with this conclusion. He explained that what was happening was the light was being absorbed by the electrons, giving them a higher energy state. And, that higher energy state was causing them to leave the atoms they had been a part of. (For this, he was awarded the 1921 Nobel Prize.) Today we use this effect to convert light into electricity in the photovoltaic cells which make up solar panels.

In 1923, an effect called Compton Scattering finally did prove that light is made of particles. Photons were made to collide with free electrons and the photons and electrons bounced off of each other like balls. Light was proven to be a particle, and also proven to be a wave, except there is no medium for it to travel through.

We began finding that other things, in fact all things, have these wave and particle properties. But, because of the lack of a medium, the wave aspect of objects is generally not considered to be real waves but a mysterious wavelike behavior.

Obviously, a known particle, such as an electron, could be shown to behave as a particle, but it could also be shown to behave as a wave. A beam of electrons could display the same interference pattern which light does. Atoms could, too.

The double slit experiment, as Claus Jonsson used in 1961 with electrons, and also helium and hydrogen atoms, is the popular test for wave properties.

In it, a beam of particles is projected at a barrier which has 2 very narrow slits in it. (This is similar to Thomas Young's 2 pinhole experiment.) Particles passing through the 2 slits would be expected to hit the screen on the other side of the barrier in either location, where the slits direct them.

But, they don't. Electrons passing through the slits create the same bands, interference pattern, which light does. Atoms and molecules do, too. Even large ones.

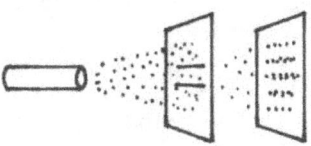

Anton Zeilinger performed the double slit experiment with carbon (C_{60} and C_{70}). Those are molecules made up of 60 or 70 atoms of carbon, each. Those large molecules were found to interfere with each other like waves.

Due to the lack of a medium for waves to travel through, the popular description of most objects is that they are particles which have wavelike properties. In fact, all things are sometimes represented to be particles, including light, nuclear forces, and gravity. While most are considered to be particles with wave-like properties, light is often considered to be a wave with particle-like

25

properties. This seems to be due to light not having mass. But, as a wave, its need for a medium goes unanswered.

In 1923-1924, Louis-Victor de Broglie introduced his wave theory of matter, with its formula of:

$$(wavelength) = h/(mass \cdot speed)$$

$$h = 6.6252x10^{-34} \text{ Joule-seconds},$$
$$\text{mass in kilograms},$$
$$\text{speed in meters/second}$$

Looking at this equation, we're dividing a very small "h" (numerator), which is Planck's constant. If we divide it by a mass of 66 kg moving at 1 meter/second the calculated wavelength is 1.0038 x 10^{-35} meters (about 1 Planck), the smallest measure of distance. This is like a 145 pound person walking casually at 2 mph or a 20 pound dog running at 16 mph. This is the smallest possible wave and it's too small to be of note.

So, it only has meaning with very small objects not moving too fast, or extremely small objects which are moving very fast. It has use in describing the movement of particles, as we see in these examples. It explains the wave properties we just discussed. Just for reference, a slow-moving grain of sand is too big and fast for you to see an oscillation.

Basically, if you can see the object moving, then you can't see its waves. The object has to be too small for you to see in order for its waves to be big enough that you can see the results of them.

Heisenberg Uncertainty Principle

In 1927, Werner Heisenberg found that we couldn't be sure where a particle is, while also knowing how fast it's moving and in which direction. This is known as the Heisenberg Uncertainty Principle. The more precisely we measure location, the less precisely we can know its momentum (speed and direction). The more precisely we measure momentum, the less precisely we can know its location.

This can be proven with a single slit experiment. By creating a narrow slit in a barrier, we can know precisely where a particle is as it passes through. The narrower the slit is, the more precisely we know that location. But, after the particle comes out of the other side, we can't be sure where it will land on a screen because we don't know its direction. The narrower the slit is, the broader the area becomes where it may land. The broader the slit is, the narrower the area becomes where it may land. This may seem wrong, but it's the true behavior of particles.

When we know precisely where the particle is, the narrow slit, we don't know which direction it was going when it passed through the slit.

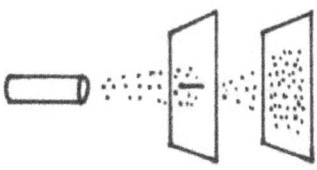

When we broaden the slit, to get a more precise landing pattern, we can determine its direction, but we've given up knowing where it is passing through the slit, its location.

You might think to just narrow the slit and send one particle through. You'll know where it was when it went through. And, seeing where it lands, you think you can tell the direction it traveled from the slit to the screen.

However, if you repeat the experiment it will land somewhere else. This is because you've affected the direction of the particle with the slit. The angle it comes out of the slit may not be the same as the angle it went into it. Because it may land in various places, we know particles don't travel straight.

Just because their destination is straight in front of them when they're unobstructed (not traveling through an opening or near an object) doesn't mean the path they take to get there is straight. They may take any path.

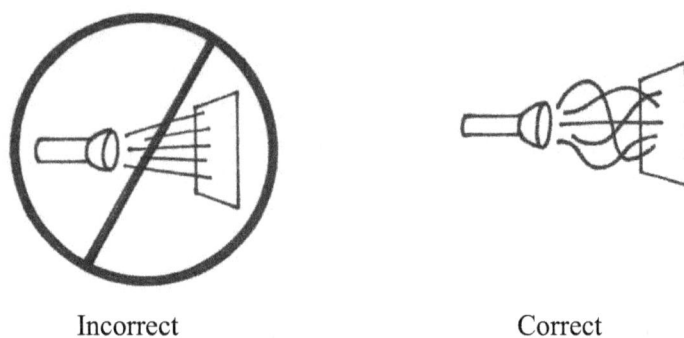

Incorrect   Correct

Superposition

Objects can be at more than one place at once. This means that particles may get in their own way.

We test for this with a double slit experiment. Remember that a beam split into 2 beams by going through 2 slits would interfere with itself on the other side of the barrier and create an interference pattern on the screen.

Obviously, being made up of many individual particles, or quanta, it was able to separate. But, a single quantum, one indivisible thing could not separate. So, it could not interfere with itself. Except that it does.

A single particle is proven to go through both slits in the barrier and interfere with itself on the other side. Instead of a beam, we send one quantum at a time, one photon (the smallest indivisible particle of light). That photon will land on the screen in one place, since it's one particle. That seems to prove nothing.

However, continuing to send one particle at a time shows us an interference pattern on the screen. The photons only land in the areas where constructive interference would put them, had they been traveling as a beam. They don't land where destructive interference would prevent them from going.

Single particles influence themselves. They can enhance or diminish themselves. This was observed in the 1970's by Akira Tonamura, Piergiorgio Merli, Giulio Pozzi, and Gianfranco Missiroli using single electrons.

When Anton Zeilinger conducted his double slit experiment, he also tried sending his large C_{60} and C_{70} carbon molecules through slits one at a time. They too, landed in an interference pattern, adjusting their location of impact by 50 times the width of a single molecule.

If you're not sure of the slit experiments, consider another way of testing for a particle to interfere with itself, the Mach-Zehnder interferometer.

This device uses a half-silvered mirror to split the wave function of a photon, standard mirrors to redirect and misalign the waves, and another half-silvered mirror to allow the waves to re-join. Two detectors are used to find the results.

I've drawn this diagram showing the wave alignment so that the interference may be seen.

The result was that every photon emitted (one at a time) only went to detector "A". Photons could not get to detector "B" because the waves coming from both directions would be misaligned and cancel each other before they got there. A photon doesn't go one way or the other at the half-silvered mirror. It goes both ways. But, since an individual photon can't split, it seems as though its probability splits and comes back together.

Quantum Paradox

Observation affects the state of matter.

It appears that the state of matter must be determined by the observer, since it doesn't seem to have any state otherwise, and can be accurately calculated to be in all states before observation.

Experiments regarding the particle/wave properties of nature always show that a thing is whatever the observer is looking for. If we look for a particle, we always find one. If we look for wave properties, we always find them.

It seems that we the observers, by looking, determine whether something is a particle or a wave. This is because we can't explain something being both at the same time. It applies to other characteristics of particles too, such as spin and polarity.

Researchers find that particles don't have a unique spin or polarity, or even location when they're created and they seem to remain this way until that property is observed. (Based on the process of creating a particle, we're unable to predict the properties of the particle.)

This is why we describe objects as wave functions. Once measured, or observed, objects seem to choose among the possibilities so we have something to measure.

This is the quantum paradox. Once we observe a single result we can no longer believe the range of probabilities still applies to our world. So, as in solving any paradox, the solution is found outside of the logical system we're using. In this case, our system is quantum mechanics.

Kurt Gödel taught us that in every logical system there are truths which the system can't explain. Quantum mechanics is a logical system and these dualities and possibilities are features which it can't explain.

In stepping out of quantum mechanics for an explanation, some people suggest consciousness collapses the probabilities. Others prefer to keep the math and speculate about parallel universes.

Copenhagen Interpretation

Neils Bohr and Werner Heisenberg led the idea that a wave function collapses when an object is observed. The bell curve of probabilities just goes away and the object is instantly and totally in the state we see it in. Observation determines the state of an object. Observation causes an object to be only a particle or wave in only one place with only one set of properties.

31

John von Neumann first explained in the 1930's that consciousness is the mechanism which collapses the wave function.

In 1967, Eugene Wigner went further and pointed out that not only is consciousness needed to collapse a wave function, but someone needs to observe the first observer. Otherwise, that observer enters the wave function. He would become in a state of multiple possibilities also.

We either create a chain of observers in which every person is part of it, or there is a cosmic consciousness, or god, which determines the properties of objects.

Many Worlds

The problem many physicists have with a collapsing wave function is that it's unscientific. Mathematics calculates very accurately the probabilities, and those probabilities don't go away mathematically. We throw them away after we observe a single state. We quit doing science and make a philosophical decision that consciousness has determined reality.

Having a problem with this inconsistency in thinking, many scientists have proposed an alternative interpretation of how we can have a super successful mathematical solution. That solution is the probability curve, which we apply to many advanced technologies, and then dismiss the unused portion as incorrect in order to rationalize what we see, and what we don't see.

In 1957, Hugh Everett suggested that instead of collapsing, the wave function (probabilities) continues to exist in the form of alternate universes. Every time a decision is made, an observation is made, or a possibility is somehow selected from a range of options and turned into a reality, the universe splits. From this line of thought comes many explanations of parallel universes and the possibilities they create.

Steven Weinberg proposed that all of these worlds exist in the same place as ours but are "decohered." Their states are separated and we are "tuned" to just one of them. This is similar to how we are surrounded by radio waves of many different frequencies but tune our radio receiver to just one of them. In the same way, we may be able to change universes by changing our tuning. This is known as "quantum jumping" or "sliding" and has been portrayed in popular science fiction."

Entanglement

In 1925, Wolfgang Pauli explained, through his exclusion principle, that there are pairs of subatomic particles which have opposite matching spin, and they are somehow connected.

Because of this, if one particle is altered the other should also be affected, no matter how far away it is. A pair can be created when 2 photons are emitted by the same electron at the same time.

In 1997, researchers at the University of Geneva showed that they could interfere with a photon there, and detect a response by another photon 7 miles away. These 2 photons were said to be a pair. If a property of one changes, the same property of the other changes also.

Entanglement is similar. It was first suggested by Einstein, Boris Podolsky, and Nathan Rosen and describes how entire groups of particles could assume matching characteristics by being in close proximity.

Roger Penrose suggests that when one of a particle's properties is altered it doesn't immediately alter the state of another particle which is paired or entangled with it. Instead, it goes back in time to when they became paired or entangled, and changes it then.

A good example of entangled particles is in stimulated emission of photons. In 1917, Einstein found that there are 2 ways in which an energetic electron can emit photons. Typically it is done by spontaneous emission in which a photon is absorbed by an electron, then one is emitted and goes in a random direction.

This is common with light sources where the incoming photon is a result of electrical flow and the outgoing photon is light. This is how standard light bulbs work.

The other way is by stimulated emission. Stimulated light is emitted in the same direction as its stimulation and the photons created are matching. They are aligned. This process is frequently applied in Light Amplification by Stimulated Emission of Radiation (L.A.S.E.R.).

It remains a mystery how particles which have this type of relationship seem to be able to communicate at any distance instantly.

Quantum Tunneling

Physics predicts that particles may violate the laws of physics.

Quantum theory has led scientists to the conclusion that not only could a particle be found anywhere, but it may even violate the laws of physics to get there. This is known as quantum tunneling. It is what happens when atoms decay, such as in the radioactive decay of materials like uranium, thorium, and potassium. In this event a neutron leaves the nucleus.

The strongest force known in our world is the strong nuclear force that holds the nucleus (center) of an atom together. But sometimes a neutron (part of the nucleus), will just leave. When it leaves, the force which was holding it attached to the nucleus is no longer needed. So the force also leaves, in the form of a very high energy, dangerous, radiation. We call this "tunneling" because it is escaping from somewhere it shouldn't be able to escape from.

Spontaneous Particles

Particles can and do spontaneously appear and disappear. This is predicted by our uncertainties of location and speed, as explained by the Heisenberg Uncertainty Principle. Just like we're uncertain of where a particle is, we're also uncertain of whether there is or isn't a particle at all.

In 1948, Heinrik Casimir observed this phenomena. He watched a vacuum between two uncharged parallel metal plates and saw particle/anti-particle pairs appear and disappear very quickly. They lasted 10^{-24} to 10^{-7} seconds each. This is known as the "Casimir Effect".

What this really is, is the substance of nothingness turning into particles. Nothing is actually something.

Our world is full and even a vacuum is full. While we can remove air, particles, and gases from a region of space, there's still something there. What's there is a field of negative energy. This is called dark energy and makes up 70% of the mass/energy of the universe.

Dark energy is believed to be what's behind the cosmological constant, what causes the universe to expand. This cosmological constant is a measure of the pressure of dark energy causing the expansion.

So there is something there, even when we determine that there's nothing. And it's that something, that dark energy, which is believed to be making itself apparent to us very briefly as particles. This is an example of quantum fluctuations, or the inconsistencies of our world at the smallest scales.

Chapter 4 - Movement In Space-Time

While we've come across a lot of difficult-to-explain behaviors, we've also discovered a conflict with our primary objective which is to discuss movement.

The conflict is that although science is about studying movement and our theories are about movement, movement can't occur in space-time due to it being quantized. We don't notice this because we do not perceive motion. We can't.

Before we get into the main issue of it not occurring, we'll explore why we don't perceive it.

Perception of Motion

We only think we perceive motion. We see it, we hear it, and we feel it. Perhaps we can smell or taste it. (I don't think most people consider smell or taste connected to motion.) But, we can't perceive it because we're not capable. Seeing, hearing, feeling, smelling, and tasting are our minds interpretations of what our senses detect. Our five senses, which we use to perceive the world are all quantized.

We have 6 million light receptors (cones) in each of our eyes for daytime vision, and 120 million (rods) for nighttime. To send the signals which they detect to our brain we have 1 million nerves per eye. Simply closing your eyes for a moment will let you see the graininess or pixelation of this limited number of receptors.

We have over 15,000 hair cells in each of our ears which detect different frequencies of vibration. We have many sensors for touch, numerous detectors in our nose, and 2000 to 8000 taste buds.

Each of these sensors sends signals to our brain by way of neurons which fire or don't fire. Like one's and zero's of a computer, our nervous system is not just quantized, it's binary.

37

There is no continuum among our senses or in any aspect of them.

Our brain is made of 100 billion neurons with a quadrillion connections. That's a lot, but again, a limited quantity, and firing or not firing.

A chess board is quantized. A piece can only be in one square at a time. While we observe a piece move from one square to another, according to chess it doesn't. According to chess, it is in one space at one moment, and in another space at another moment. Because it's quantized. Just like the chess board only knows change, not movement, our senses also only know change, not movement.

Why is it that we think we sense movement, and how then does change occur? Something's happening. Objects are different all the time. If objects get to be different, then shouldn't there have been movement?

The reason why we think we're perceiving motion is because we're detecting very small amounts of change over very small amounts of time. This is similar to a motion picture. Like old films or new digital video, we view a series of still pictures quickly.

When each picture is replaced with a new different one fast enough, we imagine that the picture is moving even though it's not. No matter how high resolution our videos get or how quickly they change, they will always be a series of still pictures, never moving. Likewise with all of our senses.

The dimensions we live in, time and space, are measurement dimensions. They are how we know where and when we are. I believe our "measurements" are true, and that they are the "judge of the validity of nature" which Feynman referred to. But, that's all we get from them. Measurement. And perhaps nothing else.

What's left is movement. Movement is real. But, we don't perceive it. So it doesn't need to occur in our measurement dimensions where our perceptions take place. And, I suggest it doesn't.

38

Motion In Quantized Dimensions

Quantized dimensions require everything to be located in a selection of specific locations. Because there is a smallest measure of distance, each location has a fixed size.

This is very similar to a chess board, but in 3 dimensions, a 3 dimensional grid. And nothing then, can be partially in one grid space and partially in another. (Remember that this is what keeps electrons in their orbits, so atoms don't collapse.)

So, for a thing to get from one place to the next, it could never exist partially there. It could not spend any amount of time partially in any space. A thing is in one space, then it's in the next. It would have to move quickly to get there in zero time so that it's never in-between. No matter how small the distance is which it needs to move, its speed would have to be infinite to accomplish that in zero time. Infinite speed is not allowed because no object can move through space faster than light.

But time is quantized also. And the smallest measure of time is the time it takes for light to move the smallest measure of distance. Each measurement is referred to as a Planck, which we discussed previously.

In the long run, we're not breaking the speed limit if we move one Planck of distance in one Planck of time, the speed of light. But, we must still move instantly in these steps. There is no transition. As Niels Bohr explained, electrons do not pass through the space between orbits when electrons change orbits, no matter how fast.

When an electron absorbs a photon, that raises its energy, which causes the electron to move to an orbit further from the nucleus. But, in getting to that new orbit, the electron does not ever occupy space between the old and new orbits.

It changes location instantly and is considered to be breaking the speed of light. (Quantum tunneling is what allows this.) Getting from one place to another is like seeing dancers under a strobe light.

The world blinks with every Planck of time and anything that wants to change location may do so. But, that's not movement. It's teleportation. Are we teleporting?

Relativity describes movement, not teleportation. So, how can relativity hold? Movement requires a continuum as Noether's theorem explains.

Movement involves infinitely small increments of distance occupied during infinitely small increments of time.

How We Move

Here is my model for assembling what we've learned.

When I put together relativity and quantum theory I get a structure for how changes in location occur, which accounts for varied experience of time and distance.

While I haven't seen movement explained in this way, this seems to be the model which prevailing theories and their proven results support. It's what I use to help myself understand particles moving, or relocating, at the quantum scale.

Keep in mind that this scale is extremely small. One Planck is only 10^{-35} meters in distance or 10^{-43} seconds in time. Even a typical atom, which is too small to see with an optical microscope, is 0.0000000008 meters wide. This would be 8×10^{25} Plancks. (That's 80,000,000,000,000,000,000,000,000 Plancks or 80 thousand billion billion Plancks wide.)

In chess, movement occurs in the air above the board. No grid, no squares, just a big hand picking up a small statue and putting it back down in a different space. It does so according to a set of rules.

To understand how our movement occurs, we need to look at space-time as a grid, like a 4 dimensional chessboard (3 dimensions of space and 1 dimension of time), since it's quantized, and has rules.

For purposes of showing it simply, and allowing me to draw it in 2 dimensions, one direction will represent time and the other will represent space. All 3 dimensions of space will be represented by one dimension on our grid. This way our grid looks like a chess board which goes on very far in all directions.

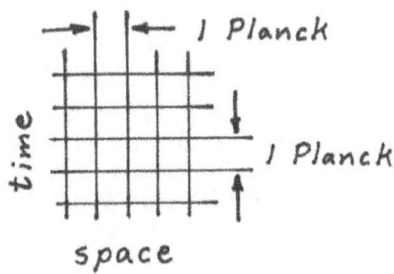

Our unit of measure in the time dimension will be Plancks of time. This I'll display as up and down (y-axis). Our space dimension will be side-to-side (x-axis) in Plancks of distance.

We'll start with 2 rules:

1. Everything must move forward in time together.

2. Any movement in space must be no more than one Planck of distance for one Planck of advancement in time.

Within these rules, everything continues to co-exist in time and not get lost or disappear from our experience of the world. They don't fall behind in time.

The fastest a thing can move through space-time is diagonally, 1 Planck of time forward (y-axis) and 1 Planck of distance to the side (x-axis). This is the speed of light. And, the change in location occurs by hopping, teleporting according to a set of rules. In this way, the relocating object is giving up experience of time and distance in exchange for relocating. Light then, is constantly hopping diagonally. It never remains to experience time or space.

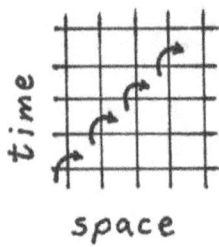

space

Something moving at less than the maximum speed (the speed of light), is periodically hopping, exchanging experience for relocation.

While at the maximum speed, it costs a whole Planck of experience of time and distance for every Planck of distance gained. Hopping less frequently, the exchange rate is better. This exchange rate is defined by the time and space dilation formulas we used earlier.

In my example in a previous chapter, I was on a rocket traveling at half the speed of light. For this rate of change in location, I gave up 13% of my experience of time and distance. I only experienced 1.73 hours during my 2 hours of traveling and 606 million miles of distance during my 700 million miles of relocating.

The exchange rate of trading experience for change in location is not a simple ratio. It is biased toward slow change. This is why we normally don't notice our loss of time or distance experience.

It's only at very high rates of change (high speeds), such as those of subatomic particles and cosmic bodies that it really has an effect.

We can see the exchange rates on this chart.

TIME & SPACE EXCHANGE RATES			
Speed	Experience Given Up	Planks of Experience Given Up	Planks of Distance Gained
670,616,629 mph	100%	1 for	1
600,000,000 mph	55.5%	1 for	1.8
500,000,000 mph	33.4%	1 for	3
400,000,000 mph	19.8%	1 for	5
300,000000 mph	10.6%	1 for	9.4
200,000,000 mph	4.6%	1 for	21.7
100,000,000 mph	1.1%	1 for	91
1,000,000 mph	0.0000011%	1 for	91,000,000

This seems to suggest a difference between matter and energy.

In this model, objects leave the quantized grid of space and time to relocate. It's between leaving and coming back that movement takes place. And I believe it's that act which may give an object the qualities of being matter or energy, a particle or a wave.

A thing which is hopping is energy. A thing which is not, remains in space-time and is matter. Most objects are both.

In their case, they alternate between the 2 states. They are energy while they hop and are matter while they are in-between hops.

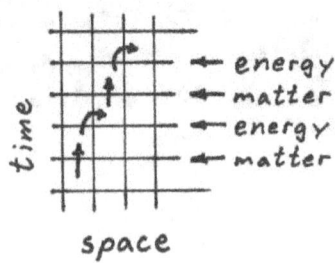

Constant hopping results in an object remaining as energy and traveling at the speed of light. While it visits space-time frequently, it doesn't stay there, so it doesn't experience time or distance. Not hopping it remains in space-time as matter.

Energy level is proportional to frequency of waves. The more frequently waves occur, the greater an object's energy is.

In this model, high energy/high frequency is seen as frequent hopping. We know that a particle which is not moving has no energy, only mass. When it moves, it has only energy. Over time, if it is to have mass, it can never reach the speed of light. To have mass, it needs to regularly rest in space-time.

This satisfies our experience of particles and waves. Particles have a location and size. Waves, or energy, have strength. It also accounts for varied experience of space-time.

Simply, energy of movement may not be able to experience time or distance, but mass must experience time and distance. And hopping may be the mechanism which facilitates this.

Wavelength is inversely proportional to energy and frequency. Longer wavelength is seen as longer periods between hops.

What determines which of these is occurring is how fast the object is moving through space.

At the speed of light, nothing can spend time in between hops sitting in space. At the speed of light, longer wavelengths mean longer hops between spacial locations, less appearances in space-time.

A slower moving thing, matter, is sitting in space. So its longer wavelength is a result of being in space-time for longer periods between hops if it has high mass, or longer hops out of space-time if it has low mass.

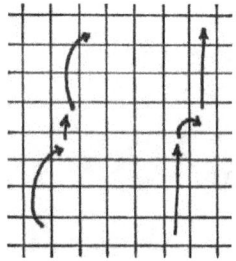

We've accounted for how things get around in quantized space-time according to the rules for relocation. But, we're still teleporting, not moving, because when we leave one space on the grid we instantly appear on another. There are no transitions on a grid.

Like chess, we've been leaving the grid to move. We've been leaving our world of space and time, as quantum theory seems to require. But we still need real movement for the laws of physics to truly apply.

A New Dimension

I propose a new dimension to explain where objects go when they hop. This allows for change of location and wave properties. It solves the need to leave space and time in order to have movement. It gives us a "movement" dimension. And, it explains what appear to be violations of the cosmic speed limit. Einstein said no object can move through space faster than light. Electrons changing orbit don't move through space.

We now have something which is leaving space-time and coming back to it. It seems to be matter (a particle occupying a specific location) while there, and energy while not there, since it has no location then as a particle should have.

It could be the wave aspect of that particle while it's relocating, but has no location then so it can only be energy.

So what is the wave traveling through in this movement dimension?

A New (Old) Substance

To solve this, I propose what may be considered a new substance. Perhaps it's the previously believed in ether, but not as it was expected to exist. Instead of something for waves to travel through, it's something for them to be made of.

I'm going to go back to our quantum paradox here and make use of the probability cloud. Remember that the probability cloud was believed to be only a mathematical structure used for calculations, then either disposed of or split into multiple other worlds after seeing the world we live in.

Our options for rationalizing the scientific predictions of particles and our observed world had been to either collapse the wave function, the mathematical structure of the probability cloud, throwing it away, or to declare the existence of a near infinite number of parallel universes to maintain scientific integrity.

I have another proposal. This one eliminates the paradox and maintains scientific integrity, but without hidden worlds. It simply re-interprets what the mathematical structure, which we refer to as the wave function, is.

This new substance, which constitutes the particles and waves is the probability cloud which makes up particles. I suggest that not only are the waves of matter real, but the probability clouds are also. And, all of them together are what was previously referred to as the ether.

These clouds carry energy from place to place. And, the appearance of them in space-time is what matter is. It's possible that Max Born discovered the fundamental substance with his wave function, Werner Heisenberg described it with his uncertainty principle, and the single slit experiment proved its nature.

This substance, the ether, is unusual compared to materials we're familiar with. It doesn't match all the properties of a liquid or gas, but I don't think it needs to. Being what matter and energy is made of, I think it's not only allowed to be unique, it's expected to be.

Where is the ether?

When Albert Michelson and Edward Morley used their interferometer to try to detect a medium which energy might travel through, they made the assumption that their device must be moving through this medium at some rate of speed in some direction.

It's a reasonably safe assumption that since our planet rotates on its axis and orbits the sun, and our solar system travels through the universe, that an ether which spans all of space could not be still in relation to us, if it was separate from us.

In response, I suggest that the ether is us, and that is why we can't detect our motion relative to it. It is also light, which is why it can't be found to move differently than light does.

Because this substance is what makes up all particles, it also goes where the particles go. It is the clouds of probability which move relative to each other. But, the ether has no speed since it is not in the measurement dimensions. So, the reason light measures the same speed according to every viewer is that its particles move at that same speed in relation to that viewer. Its particles actually get on the plane with us when we fly. And the light we see adjusts its wavelength according to each viewer's relationship with it so that it may be experienced as the same speed for everyone.

Relationship of the ether to space-time

Since we've found that space-time is a grid, then the movement dimension must contain a large body for carrying waves. These waves are not lines or traveling particles. They are broad and span the universe as Born described.

Each wave is what we currently refer to as a wave function or probability cloud. It reaches everywhere, but has a highest probability, a peak.

The grid of space-time may be imagined to be like a screen suspended over the surface of the ether. The peaks of the waves touch the screen. They are present in space-time there. They are particles. Where they don't touch they're not present in space-time but still exist in the movement dimension.

Particles are an emergent phenomena of the ether interacting with space-time.

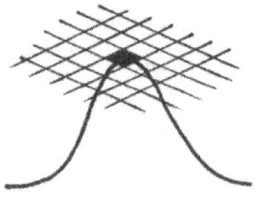

The ether is composed of all probability clouds existing together. Each is unique but can influence each other as waves do, and according to a set of rules. They may interfere as they pass through each other, creating complex waves or blocking each other.

We can see the effects in space-time as peaks (particles) interact. But their whole bodies can interact also, in the movement dimension. The results of which may be insignificant and may not appear in space-time. However, this interaction which does not extend into space-time may still affect what happens in space-time.

Chapter 5 - Theory Interpretations

Earlier I described specific behaviors or qualities which relativity and quantum theories predict, and which reliably prove to be true. I will now proceed to address each of them and how they can make sense in the world we're familiar with based on the model I've proposed for the movement dimension.

Relativity

The speed of light is constant and time and distance adjust themselves to accommodate that.

It's my belief that the speed of light is fixed as a result of distance and time having fixed minimum measurements.

In a grid, when moving forward and sideways at once, through time and space, the only available grid-space is diagonal. Constantly advancing in time and distance is the inherent speed limit for moving in a space-time grid. So, the grid structure creates the speed limit as being the speed of pure energy, an object with no mass (no experience of space or time).

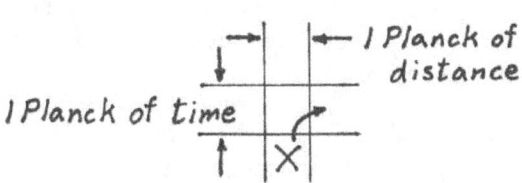

Distance and time being quantized brings us a demand for a movement dimension. Because of the need for particles to leave space-time in order to relocate to other places, some loss of their experience of space and time is expected.

If a particle had not left time and space, but had just moved slower through time and space, then that particle wouldn't be where or when we are. It would be behind us in history until it caught up. Otherwise it would no longer be a part of our world. Time and distance becoming lesser only places things in different locations in space time.

Leaving the grid and coming back allows an object to be with us, but skip some of the experience of time and distance which we had. It explains how two objects which were together may separate and come back together having experienced different amounts of space and time while they were apart.

Mass and energy are equivalent.

Most physicists today believe that all objects are particles which have wave-like properties. And many would even say that the wave aspect of matter doesn't represent reality, that there are no real waves.

I believe this is due to their appearance from the measurement dimensions of space and time. Because waves don't occur in space and time they can't be seen there, only their results can.

I propose that the waves and particles are both equally real representations of the underlying substance. They emerge from it. And, this substance is the probability cloud, which is real and exists in the movement dimension.

The particle and wave properties are a result of this cloud interacting with the grid of space-time. Matter is the existence of the cloud in space-time as a particle. Energy is the movement of the cloud in the movement dimension. Because of this, everything can be described as a wave and a particle, and the energy/matter equivalency is due to them being two different expressions of the same thing.

In this model, the ether (all of the probability clouds) is the fundamental substance of all things. The waves are real properties of the ether. They are energy. And, the ether's presence in space and time creates the particle property, which is matter.

Both matter and energy, particles and waves, are equally real. And they're equivalent because through motion, objects regularly transition from one to the other fully.

This full transition, the fact that when they're in space-time they are entirely matter and when not in space-time are entirely energy, is what truly equates them. They emerge from the way the cloud, or ether, interacts with space-time.

Gravity is not a force but a warping of space-time.

I don't think the warping of space-time needs to occur, but don't believe I could detect it if it did, since everything I might use to measure it would also warp. I believe there could be other explanations for the effects of gravity which may be similar to the mechanisms of other forces.

Because gravity is proportional to mass, it is also proportional to energy. But energy does not have to move in an obvious way.

We know that particles not traveling are still moving. When the mass/energy is high enough, I suspect its movement could become significant.

My suspicion is that the wave aspect of massive bodies has an influence on nearby objects. And, because of intense wave activity, these objects are not remaining in space-time.

The movement involved in wave activity could account for some absence from space-time, some lost experience of time, as traveling does. And the gravitons which many scientists believe communicate gravity may be a part of this activity.

Regarding gravitational waves and their detection. I believe if a change occurred in the structure of space-time, a warp or ripple, we wouldn't be able to detect it with instruments which are also located in that warped or rippled space-time. The instruments would be altered to fit space-time. So, I suggest that the gravitational wave which was detected, was not part of the structure of space-time, but was something else.

A continuum is needed for physics to be consistent.

Earlier, we saw that Noether's theorem requires a continuum in order for rules to be obeyed. Our dimensions of space and time are quantized. They have minimum divisions and cannot be divided further than those.

Since they're not infinitely divisible, they're not continuums. They're not smooth. They seem smooth because these smallest measures are extremely small. Even at the atomic scale, they would seem smooth. A typical atom is 8×10^{27} Plancks in diameter.

Since they're not continuums though, they can't satisfy the requirements of Noether's theorem.

Our movement dimension, as I propose it, is a continuum. It is not quantized. And, it is involved in all movement. True movement, as relativity specifies. It interacts with all of space-time through its ether.

The movement dimension with its ether provides the continuum and true movement which makes our laws reliable. And, the quantization of space-time accounts for violations of these laws, the inconsistencies in the behavior of small particles as they are forced into a limited number of locations.

Quantum Theory

Our world is quantized.

Quantum theory explains that space and time have a minimum quantity of measure. And energy does also. This quantization is necessary for our world to function the way it does, for atoms to behave properly.

We've explored how the continuum of the movement dimension and its probability clouds can fit together with the quantization of space-time. We see the necessity of quantization for providing the effects we experience, such as fixed electron orbits and the ability for objects to behave as both waves and particles.

It seems that the ether (all the probability clouds) interacting with this quantization is what makes our world what it is. The grid is the rules which the substance obeys.

Things exist as probability clouds or wave functions.

I propose that the probability cloud, or wave function of a particle, is the ether. I suggest its structure according to Max Born should be interpreted not only as a representation of a particle's probability, but as a whole body in the shape of the calculated probability.

In this model of movement, the probability cloud is the fundamental substance which presents itself as matter when located in space-time. And, energy is the waves in the cloud as it relocates in space-time.

So, I suggest that not only do objects exist as probability clouds, but these clouds are the most basic component of all things. Particles are the way these clouds show themselves in space-time and waves are the way they behave in the movement dimension.

Everything is a particle and a wave.

As we see in this model of a special dimension to accommodate movement, this is absolutely true and necessary. The particle and wave aspects are real and are not just descriptions of the behavior of objects. They are a result of the substance of things interacting with the grid of space-time. This is similar to the mass/energy equivalency.

Based on the inability of movement to occur in our quantized space and time dimensions, and the necessity of leaving these dimensions in order to relocate, everything is both a particle and a wave. They are particles while they exist in space-time. And while they're relocating, actively moving in the movement dimension, they are waves.

Since the ether (the probability cloud of a thing) is our fundamental substance in this model, it is the ether which presents itself as a particle in space-time and forms a wave as it moves.

Uncertainty of location or movement.

Werner Heisenberg explained that the more precisely we know the location of a particle, the less precisely we know its momentum (speed and direction). In the single slit experiment, a narrow slit led to unpredictable results in direction once an object passes through it. A wide slit showed more predictable results.

If we look at our object as a probability cloud, not just its particle (peak), then we see a more sizeable object squeezing through a narrow space. This is because the cloud can have no probability of existing where the barrier is. Its only probabilities can be on either side of the barrier or in the slit area. So, coming out the other side, the probability spreads suddenly.

Since experiments tell us it spreads very broad, it must get extra fat on the far side of the narrow slit. The narrower the slit, the more the cloud seems to pop out the other side. The wider the slit, the more the cloud seems to maintain its natural free moving figure.

The location of a particle and its momentum emerge from the behavior of the fundamental substance. Our uncertainty comes from the way it interacts with the substance of other objects and also the grid.

Superposition – Things can be at more than one place at once.

As we are used to thinking in our world, with only our 3 spacial dimensions and a time dimension, being at more than one place at a time is hard to imagine. But, with the existence of a movement dimension and with ether (the probability clouds) being the substance of our world, I believe this is easier to accept.

We currently use the probability cloud as a model for calculation purposes only. But, if it's real, then it becomes easy to understand superposition. While the particle is only located in space at one place, its cloud exists all through the movement dimension to some degree. Remember that the location in space-time is only the peak of the cloud. This allows it to affect objects which are located anywhere.

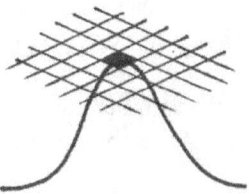

To go further, superposition, and especially evidence of particles interfering with themselves, suggest that the underlying substance of a particle may change shapes.

The Mach-Zehnder interferometer experiment we looked at earlier describes a wave function splitting into two parts and following two different paths simultaneously.

Expecting that an object is represented by one particle makes the phenomena seem unlikely. However, thinking of the object as a real cloud not just a probability calculation for a single particle, it's easy to imagine it splitting in two and rejoining or just spreading to occupy all possible paths.

That doesn't mean an indivisible particle divides, only that the substance it represents spans a larger area than we observe and its shape changes. It may have multiple areas of concentration. Some of the probability cloud divides.

Quantum Paradox

Consciousness affects the state of matter. Or, we choose which parallel universe we live in.

I suggest that consciousness doesn't actually affect the state of matter. It only appears to. And, that parallel universes aren't actually implied by quantum theory.

If movement occurs in its own dimension as I've described, then matter is regularly and quickly transitioning to and from energy, between particle and wave. It can almost be considered to be both at all times because the transition is so frequent in order to appear as smooth movement to us. (The rate of this transition would be calculated by de Broglie's formula for the wave theory of matter.) The underlying substance of the object remains the same.

What we experience, what we can measure in our dimensions of space and time, is always the particle aspect of matter. Our knowledge of the wave quality of matter is derived from the way it affects its particle position and the positions of other particles. Where we find particles to be, how we find them arranged, tells us of the waves which got them there.

Regarding the collapse of a wave function, the elimination of possibilities due to observation, I don't think it needs to occur.

Properties such as spin or polarization may be set at creation and variable characteristics like momentum and location may be constantly changing, in which case the probabilities always remain, only the peak changes.

When a measurement is taken, when a particle is found to be at a specific location at a specific time, for an instant we say its probability of being somewhere else becomes zero. This is true.

However, in the instant before that measurement and in the instant after, the other probabilities are there. It's only for the smallest measure of time that they're not there. Even so, the body of the cloud may remain.

Entanglement and paired particles.

These two phenomena may be different from each other, although they may sometimes appear to be the same.

Paired particles may be two different peaks of the same probability cloud. They may be just as connected as if they were the same particle. And due to the infinite span of their probability cloud, they may also be infinitely separated.

Entanglement may be when two or more separate clouds take on each other's qualities or even become one cloud. Because their probability concentration is highest near their peak, they are most able to influence each other there and take on each other's qualities. At a distance, low probability areas of the clouds have less influence and are more free to be unique from other clouds.

Imagine you're walking with a friend. You'll want to be at the same speed, but don't need to be in step or even of the same stride length. Now get close. The closer you get, whether it be behind, beside, holding hands, or with arms around each other, the more you have to match your movements. You'll end up in perfect alignment if you walk together close enough.

Regarding Penrose's suggestion that an entangled particle may go back in time to affect its entanglement partner. I agree that it could because the ether which presents itself as the particle is not in space or time dimensions.

So, it can have influence throughout all of time and space. While its peak may only exist at one location in space-time, its body (probability cloud) spans all of space and time, including the past and future.

Tunneling – Violating physics.

Tunneling, described as the behavior of a probability cloud, is when a particle appears in space-time in a low probability area, having just been in a high probability area.

When the move occurs, the probability peak relocates. This would be as a wave peak drops down and re-appears somewhere else. As with any time a particle relocates, its probability peak relocates with it. The probability behaves the same as the particle.

In the case of radioactive decay, the peak drops down and relocates. Why it occurs seems to be due to its intense concentration.

A very fine peak, eventually, can't hold anymore and falls. This is a way of visualizing the Heisenberg uncertainty principle in action and is very similar to a wave squeezing through a narrow slit. Very precise location can't maintain itself.

This wave can persist.

This wave cannot persist because it is unstable and will collapse.

This would be why we see it occurring in nuclei of large atoms where density is high, mass is concentrated and confined. And just as it was hard for a particle to get out, it's also hard for it to get back in, so it stays out.

If you picture a wave function (the bell curve) as a gelatin dessert, you can imagine the wobbling of the top of the bell as quantum fluctuations. The taller and narrower your dessert is, the more likely its top will move erratically.

Or, to be more comparable to one of our experiments, since we can't watch a particle move, the taller and narrower your dessert is the less you can predict where its peak will be when you photograph it.

Spontaneous appearance of particles.

Spontaneous appearance, described in terms of a real probability cloud, is when a spread out (low) probability develops into a concentration, a high probability. There could be various reasons for this to occur.

It could be the result of the peak concentration of a probability collapsing somewhere else and appearing in view, as in tunneling. And we just didn't notice where it disappeared from.

It may be an accumulation of very small concentrations passing by each other, which temporarily create a composite peak. If multiple small concentrations temporarily assemble, then they could just as quickly disperse, retaining their original properties.

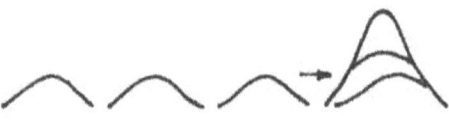

Maybe the particle didn't really appear spontaneously. It might be a low frequency, long wavelength particle passing through the area. A broader view of the area may reveal multiple appearances of the same traveling particle.

Chapter 6 - Particle Properties

In the world which we experience, simple movement through space and time seems to describe most events we're familiar with. But, we know there's more to physics than that due to the obvious force of gravity and well known effects of electricity and magnetism.

Examining our world closely we find a diverse array of particles which seem to be fundamental and have unique properties until they get broken apart into smaller ones with other properties.

As we dig down through the layers of science we arrive at atoms, made simply of a nucleus orbited by electrons and exchanging energy in the form of photons. Variations of these parts seem to give us all the emergent properties of our chemical elements and lead to all of the higher layers of science.

The atomic structure seems fairly simple. But, when broken apart, the pieces of the atom as well as the many other particles existing at smaller scales are much more complex. What we find are particles and fields. And, they don't seem to be fully described by their movement, their relationship with the space and time dimensions.

They have properties such as electric charge, spin (which doesn't seem to be physical spin), and colors (which are not visible colors). Actually, they have relationships which we assign creative labels to in order to help us understand how they interact.

In our search for the fundamental makeup of our world, we find many particles which seem to be fundamental. But, we also find that they are all made of the same substance. The same substance somehow having different descriptions.

We know that all these particles are made of the same substance because we can make them using other particles. It doesn't seem to matter what type of particles we use, we can still make all the others. This is what we do with particle accelerators.

The common explanation for how this works is that by colliding particles in a field we're able to knock loose some of the field in the form of new particles. However, the mass of what's found after the collision matches the particles and energy which collided. In addition, as the new particles decay, they turn into yet other particles. Perhaps if something is being knocked loose from a field it is the properties of the field. I suspect we're just describing the same phenomena in different ways.

Currently, our most powerful particle accelerator (the Large Hadron Collider in Switzerland) uses only protons. By colliding identical protons at high speed, 99.9999992% the speed of light, we can create various other particles. (High speeds are reached by contributing energy to the protons in the form of photons which become part of the protons.)

Once the protons collide, the shower of newly created particles is separated and they are identified by their properties such as mass, energy, and electric charge.

Some accelerators work by colliding electrons in a similar manner. They too, are accelerated (given energy) in the form of photons and collide to produce various new particles.

Some of what's produced in particle accelerators are particles commonly found in nature, like electrons, photons, muons, and neutrinos as well as their anti-particles (which are common too).

Other particles made in the collision are not found outside of particle accelerators, such as the Higgs boson, quarks and gluons which are normally bound so tightly they're unidentifiable, and exotic combinations of quarks.

As we continue to collide particles we occasionally identify new ones. Often the existence of these new findings are predicted, as with the 2012 identification of the Higgs boson.

The Higgs boson was predicted in 1964 by Peter Higgs, Robert Brout, and Francois Englert. It was initially proposed by Higgs who found it a necessary part of particle theory. Though it's uncertain how it works, the Higgs particle and the field it emerges from is believed to create the phenomena of inertia, the mass aspect of objects. Without it, theories predict all particles to move at the speed of light.

Sometimes unexpected particles are discovered, such as the muon. But, any particles which are not a normal part of our world quickly decay into particles which are. They transform. They break apart and acquire properties which make them common to our world.

Since we can make all these particles using whatever we have which is convenient, we know that their fundamental substance must be the same. We saw this suggested, also, through our exploration of movement. So, our puzzle now is to learn why there are different particles, why our fundamental substance presents itself only in specific ways, not in every way imaginable.

At this point we can say that perhaps there aren't different particles since they are made of the same fundamental substance. Only their description is different, and their description is the key to their uniqueness. Specifically, the descriptors are the key. But they are something we can change only through great effort.

In our study of movement, we found that the fundamental laws of our world seem to be about the way this substance interacts with the grid. The grid referred to the quantized dimensions of space and time.

The fundamental laws of our world seemed to be regarding the interaction of the substance with the dimensions. Through this we saw how matter and energy are different from each other.

Looking only at matter and energy, just two types of substance, we were able to describe their behaviors using the dimensions of space and time. Space and time are the descriptors which make matter and energy what they are, unique from each other. They are also the descriptors which we experience with great richness in our world.

Matter and energy appear to be defined by their changes of location in space and time. In addition, the wide variety of particles we find or create are also defined by their descriptors.

The dimensions of space and time help to describe the various particles by way of their movement. But, other descriptors are needed as well in order to explain all of their uniqueness.

Particles have properties other than velocity and location. These other descriptors are not all obvious parts of our experience of the world, but do contribute to the variety of materials we have and their behaviors.

Some of the additional descriptors we need are:

- total mass/energy

- electrical charge

- spin

- generation

- chirality (direction of spin)

- color

We assign these descriptors because we need ways of describing aspects of particles which we haven't been able to explain simply by their location or changes in space and time.

These descriptors are given somewhat arbitrary names which are not what their words imply. They don't refer to a physical spin, visible color, or particles giving birth to other particles. They are used to explain relationships and interactions between particles. Just like particles described to have matching locations (or distance quantities) are caused to change direction by colliding spacially, particles can also be caused to change directions by attracting or repelling each other based on their electric charge, spin, or color quantities.

Descriptors

Imagine a dog which is 3 years old, 24 inches tall, has long hair, is at the park, and is running west at 10 mph.

I've told you of the uniqueness of the dog using the descriptors of age, height, hair length, location, and velocity.

These descriptors can be reduced as phenomena of other descriptors. The dog's age is a phenomena emerging from his relationship with time. His height is based on his relationship with the up and down dimension of space. Long hair can be described by spacial dimensions. Location is due to quantities in the spacial dimensions. And velocity is based on space and time.

In this example, I was able to use only space and time to describe all of these qualities of a dog. Space and time are the descriptors we're most familiar with. Much of our experience of the world arises from us and other objects changing quantities in space and time.

This is what we do in physics. We reduce objects to their most fundamental descriptors and study relationships with them, often how objects interact with space and time.

To fully explain subatomic particles, we include descriptors which we haven't yet figured out how to condense. Sure, particles have locations, velocities, and behaviors we can describe using space and time. But they also have unique qualities which don't seem to emerge from their relationship with space and time.

For example, if we want to describe an electron we would tell its up or down location in space, side-to-side location in space, forward or backward location in space, location in time, electric charge, energy/mass, spin, and chirality (direction of spin).

To describe a quark, we would tell its up or down location in space, side-to-side location in space, forward or backward location in space, location in time, electric charge, energy/mass, generation, and color.

We would need 8 descriptors for an electron and a slightly different 8 descriptors for a quark. But, with these 8 descriptors we can tell all there seems to be about these particles.

There are no known differences between an electron and a quark other than these qualities. Aside from them, electrons and quarks, as well as all other particles, seem to be the same. And, since they can be changed into each other they seem to be made of the same substance.

Unlike my previous description of a dog, we're unable to consolidate descriptions of electrons or quarks to only space and time. (This is partly due to how much I didn't say about the dog because I wanted to make a point about the unique properties that particles have.)

This seems to leave us with 4 dimensions and several other descriptors sitting at the foundation of science. We can describe anything in our world using only a relatively small number of descriptors.

The total number of descriptors may be as high as 40 today in order to encompass every known particle (there are hundreds) and their properties. I suspect this number can be reduced as we learn more and are able to consolidate them like we have with all the qualities we can fit into space and time.

Now let's consider what these dimensions and other descriptors really are.

Dimensions

How far is a foot? It's a foot. It's 12 inches. One-third of a yard. I can only describe distance as distance.

What if what we call a "foot", is actually only an inch? Does that mean I'm shorter than I think I am?

In 1931 James Jeans suggested that the universe isn't expanding but instead we're shrinking. Is that really any different from saying that the universe is expanding? If I and everything around me is shrinking, including my ruler, how would I know? Does it even mean anything?

We can expect that in earliest times most people believed the Earth was the center of the universe. Then, in 280 B.C., Aristarchus said the sun was the center of the universe.

For a couple thousand years we went back and forth over this issue and seem to have settled on the idea that everything is in motion relative to everything else. Is anyone right? Is there a center of the universe and does the question have any meaning? Are we moving through the universe or is the universe moving around us?

When we learn to walk or ride a bike the floor or ground often hits us. Or, perhaps we are hitting the ground or floor. There isn't a difference.

We select one thing as a standard and compare everything to that, and that's all we can do. Whether a foot is big or small I don't know. I only know if it's bigger or smaller than something else. We only know of relationships.

So, I point one way and say that's the up and down dimension. I point another way and say that is forward/backward. And another is side-to-side. It's how I describe things compared to me. I do the same with time. When I say I'll be there in a minute, I mean a minute from when I said I'll be there in a minute. I don't know when I am in "all of time." I can only compare various events in time.

Our senses tell us of the world around us. They even tell us of our experience of space and time. But they really only feed us information which our mind interprets as experiences of space and time. And the experience is an emergent quality of changes based on space and time.

Through virtual reality technology, we know we can remain in one place while sensing travel. We can be fooled by feeding our senses the same information which experiencing the world does.

Einstein explained how gravity and acceleration are equivalent experiences. Gravity and wind and a changing view can also replicate travel.

This doesn't make our dimensions of space and time not real. They are real. But, they are really simply descriptors.

Through virtual reality we are fooled into thinking the descriptors are different from what they really are. But, what they are remains the same because ultimately they're based on real and reliable relationships.

Time can only be described as time, and distance can only be described as distance. They are fundamental as far as descriptors go.

When we look at the descriptors we use for basic particles, they too are dimensions unless they can be explained using other dimensions.

Chess is a game which takes place on a two dimensional board, and in one dimension of time. It's divided into 8 forward/backward locations and 8 side-to-side locations of space, and perhaps 20 to 30 moves or units of time.

Checkers, also, takes place on a two dimensional board. It's dimensionally similar to chess, except that when a piece reaches the far side of the board it gets "kinged." It gets another checker placed on top of it so it can proceed to move in reverse.

It has an extra dimension. Checkers has 8 locations forward or backward and 8 side-to-side and what might be considered 2 locations of height. That dimension of height, even though it only has 2 levels, is what's used to describe the 2 types of playing pieces: the regular checker, and the "kinged" checker.

But, since the height dimension is only used to describe the type of piece and is not accessible as a location when relocating a piece, it's not considered spacial. This makes checkers a game of two spacial dimensions, one type dimension, and one time dimension.

Now we can see we haven't fully described chess since we haven't explained all the various types of pieces it has. In chess, while we have 6 types of pieces, they're not described by their height or stacking. Even though its pieces, to us, are 3 dimensional statues, the game of chess only recognizes their type. So chess, too, is a game of two spacial dimensions, one time dimension, and one dimension of piece type (which extends to 6 possibilities). It has 4 descriptors.

We should also include the movement dimension we discovered earlier as a necessary part of accommodating change. We can add this to both games.

Our world can be described in a similar way. It has 3 dimensions of space which may extend infinitely far (have infinite possible quantities), one dimension of time which also could be infinite, one of movement which must be infinite because it is infinitely divisible in order to be a continuum, one of electric charge (with possible quantities of -1, -2/3, -1/3, 0, +1/3, +2/3, +1), one of spin (with possible quantities of 0, 1/2 or 1), and many more.

Transitioning

There's a difference between space and time and our other descriptors for particles. That difference is transitions. I know that "there" is far from "here" because of my need to pass through all that's in-between in space and over time.

That passing through of all the in-between creates a geometric relationship between "here" and "there" which we experience.

71

On my radio dial I have to pass through all the frequencies in-between while I go from one station to another. The frequencies are a descriptor. The passing through of frequencies while passing through time creates a linear experience, one dimension, traveling along a line.

An Etch-a-Sketch (hopefully you're familiar with this old toy) is very similar to a radio dial, but with 2 dimensions. I turn one dial, like on a radio, and transition from left to right with a pointer. A second dial transitions it from bottom to top. Together I have 2 descriptors of location to explore and the transition of going from one location to the other creates a geometry. I can't get from any place on the Etch-a-Sketch to any other place without passing through the locations between them.

Any way I travel with the pointer I create shapes which represent the relationship, the in-betweens of my starting and finishing locations. I can see the shapes because I'm removed from the screen.

But if I were on the screen I could still experience the shapes by sensing the transition from place to place, just like riding in a car gives me a sense of the relationship between two addresses. However, if I could get from place to place with no transition, then there would be no meaningful relationships between locations. No geometry.

The descriptors we normally think of as dimensions have geometry, relationships between their measured quantities. Must a descriptor contain a geometry in order to be considered a dimension? Remember that geometry is relationships among quantities.

If I were to step into a teleportation device and enter the spacial coordinates I want to relocate to, then suddenly appear at that location, I've removed the experience of transitioning from place to place. Because I can change to any coordinate I want to, no place is further from or closer to another. The travel, the transition, is removed. There is no more geometry. Relocating, then, is no different from changing my surroundings on a film set.

When a child picks up a chess piece from anywhere on the board and places it anywhere else on the board, they're breaking the rules. Like my teleportation machine, the child has removed relationships, geometry, from chess. Since particles may do this in space, it can't be a requirement for a dimension. Fortunately, this occurrence in space-time follows strict rules most of the time.

So, relationships among different quantities of a descriptor exist when there are rules for changing them. Rules for how to get from one quantity to another creates relationships or geometries for that descriptor. And, transitioning according to those rules makes our experience of them, which seems to be all there is to know about dimensions.

In our time and space dimensions, transitioning or passing through quantities in an order gives us the experience of space as we know it. It is what makes things feel distant. We experience this through our senses.

The transitions create our experience of time and distance. But time and space aren't emergent phenomena. Our experience of them is just as our experience of color emerges from our mind's interpretation of various energy levels of photons.

The experience of space and time transitions is so important to our experience of the world that we can get sick when any of our sensory inputs aren't in agreement. This is what motion sickness is.

Other properties of particles, such as spin, electrical charge, and color, don't create the experience of vastness which time and space dimensions do because they don't cause us to experience transitions. And, they don't extend to the great quantities that time and space do.

If there were a large number of quantities for spin and they were experienced in order over time, then there would be a vastness to spin as there is with distance. As it is, spin has very limited quantities which don't seem associated with time and may not form transitions. Spin, also, doesn't obviously impact us at our scale.

In addition, while spin is linked between different particles, it doesn't seem related to distance. This may be why particles with related spin (paired particles) are able to have very different spacial quantities and that difference doesn't affect their spin quantities.

Spin is simply a dimension which is mostly unrelated to the spacial dimensions. It is its own dimension.

Property Relationships

In describing movement, we explored the relationship between matter and energy and why photons (light) have no mass, but we grouped all matter (particles) together as being the same. Now, giving them back their unique properties in the form of additional dimensions, I would describe all particles as being the same, just with different quantities in these other (non-space-time) dimensions.

If all particles can be assigned unique properties, and these properties are the only things which make them unique, then there should be no difference in their underlying substance.

This is what we find as we convert particles into other particles. The makeup of our world is one substance with a small selection of descriptors assigned to it.

In 2007, Aston Bradley and his team relocated a beam of rubidium atoms by only moving light which carried the information of the particular rubidium atoms. He gathered the information from one collection of rubidium atoms, all of their descriptors regarding space and time (location, energy, velocity). Then he gave that information to a different collection of rubidium atoms. The second collection of rubidium became identical to how the first collection was.

The way he did this was to lower the temperature of one collection of rubidium to near absolute zero. This collection was just a generic base to work from. He added to this another collection of rubidium atoms which were then caused to cool. As the new atoms cooled they had to give up their energy in the form of photons (light).

Each photon held the exact energy description of the atom it came from. The light was transmitted through a fiber optic cable and given to a second generic base of rubidium atoms. Since they were also cooled to near absolute zero they had no unique energy properties. But, with the photons arriving in the same arrangement they had left in, the awaiting atoms which absorbed them took on the exact energy properties in the exact locations and velocities as the atoms they came from.

Did Aston Bradley teleport rubidium atoms? Since the initial collection of atoms and the final collection of atoms had the exact same properties but were not moved, maybe he did.

Similar to the way we know one particle from another, I can identify my golf ball from yours. Mine is the one off to the side near the weeds, while yours is the one between the tee and the hole. If it weren't for its location description, its spacial quantities, I wouldn't know the difference.

I could change the spacial description of my ball to be that of yours, and the spacial description of yours to be that of mine. That would make yours into mine and mine into yours. By the two balls trading places, they trade identities.

The same works with fundamental particles. By changing their quantities in a dimension or two, their identities change.

Particle Identities

Let's look at some examples of how particles change from one type to another.

To begin with, we don't normally consider a change in spacial or chronological quantity to change the identity of a particle except in certain situations like a golf game.

Space is considered to be infinite and most every particle has a unique spacial quantity which is usually changing. That would make identities too hard to keep track of. Space and time quantities only identify specific particles.

So, the difference between an up quark and a down quark is simply mass and electrical charge. Other than that, they're the same. Change mass and electric charge and they may become each other's type.

Fortunately this doesn't happen too easily, otherwise our world could become very confusing. Objects may not remain what we think they are. Rules of conservation prevent this from happening. If up and down quarks could easily turn into each other, then atoms would fall apart and our world would crumble.

Conservation just means that all quantities are saved, at least regarding the non-space-time dimensions. This means that if there's a +1 electric charge then something has to have that charge. If one particle gives it up, another one must acquire it. This exchanging or passing on of these quantities among particles is limited to certain events like radioactive decay.

When a uranium atom decays, a neutron and the gluon (energy) which held it in the nucleus leave. The neutron can't survive on its own, so it decays into a proton, electron, and anti-neutrino. Neutrons have zero electric charge, so that is conserved by the proton having a +1 charge and the electron having a -1 charge. Gluons change into photons, both are forms of energy with no additional quantities to be conserved.

Electrons can't change into "top" quarks all by themselves because they have different electric charge quantities. But, an electron and an anti-electron together can form photons. They can do this because the -1 electric charge of the electron and the +1 electric charge of the anti-electron cancel each other to make the zero electric charge photon. And, the photon can accommodate any amount of total mass/energy the electron and anti-electron had.

Mass/energy is conserved, too. High energy photons may turn into electron/anti-electron pairs. As long as it's the pair of them, then their quantities are conserved.

Anti-matter is often involved when particles change into other particles. Anti-matter helps conserve the quantities because anti-particles are like their corresponding ordinary particles but with opposite properties.

Then, only total energy or mass needs to be accounted for. This means that any large amount of energy or matter put together may convert into any type of particle. As long as it's accompanied by its anti-particle, their total quantities are conserved.

A quark and its anti-quark have opposite colors and opposite electric charges. Also, any properties which happen to be left over may possibly be shed as smaller particles to carry that quantity while the bulk of the energy or mass may change into any particle and anti-particle it likes. We see this in particle accelerators.

In a particle accelerator, we take a bunch of identical particles (because we can control them if they're identical) and add energy. We make them go faster by contributing photons which they absorb.

For example, at the Large Hadron Collider in Switzerland, protons are accelerated to 99.9999992% the speed of light by adding up to 14 TeV of energy to them. That's 14 tera-electron volts, or the energy of one electron carrying 14 trillion volts of force.

Two batches of these protons going opposite directions are made to collide. The collision produces a shower of many different types of particles which must equal the total mass/energy, charge, etc. of the collided particles. But, the resulting particles may be of any type fitting that description. Out of only protons, which are up and down quarks bound together, all others get produced. Those produced are measured for their properties by detectors.

Here's an example of what may happen in one of these collisions.

Twenty protons collide. Each proton has an electric charge of +1, for a total charge of +20. This is a result of them being made up of two up quarks (+2/3 each) and one down quark (-1/3 each). The collision is the same as colliding 40 up quarks with a total charge of +80/3 and 20 down quarks with a total charge of -20/3, plus all their mass and the energy added. Quarks also have color, but because they were all parts of protons their colors cancel each other and can be ignored.

The result is lots of mass/energy with a total charge of +20. This could be represented by any combination of particles as long as they total the same mass/energy as the 20 protons and an electric charge of +20.

Protons have a mass of 0.938 GeV/c^2 each (0.938 giga, or billion, electron volts per speed of light squared). With enough energy added to them they could combine to form a Higgs boson, which has a mass of 125 GeV/c^2. But needing to account for the electric charge of +20, we might also get 20 anti-electrons of +1 charge each and only needing 0.000511 GeV/c^2 each.

The Higgs particles aren't found to naturally occur in our world so they quickly decay and usually form bottom quarks. But, since bottom quarks have a negative electric charge (-1/3 each) an equal number of anti-bottom quarks will be created as well, having +1/3 charge each.

Quarks and their anti-quarks in the same location, however, will annihilate leaving behind pure energy in the form of high energy photons of whatever amount they need to equal the mass/energy of what was annihilated.

Meanwhile, the anti-electrons will likely absorb and emit photons occasionally until they encounter free electrons to annihilate with, leaving only photons in their place.

Fields

Fields are generally not considered to be made of particles, but can form particles. They are areas of space, in which something invisible can influence particles.

Examples are static electricity, magnetism, and gravity. Invisible forces are described as fields because no particle can be found, but there is an effect of the properties of particles. The properties are there, but the particles aren't.

Fields are what fills space. They are present in vacuums and this concept explains vacuum energy, which is commonly called dark energy as I mentioned earlier. I believe fields may be the fundamental substance interacting with some dimensions but not space. The ether here just doesn't form a particle, or peak.

Just to be sure I've discussed all of our material world, I'm going to include dark matter here, which I think may qualify as a field.

Dark matter, as indicated by its name, is suspected to exist as particles. But, they have either not yet been found or not yet been identified.

We know of its existence because of its gravitational influences. And, we know it's significant because these gravitational influences tell us that dark matter makes up 90% of all matter in the universe.

So, while we expect to someday identify the particles making up dark matter and we've already included gravitational fields in our discussion of the components of the universe, I wanted to mention it specifically because of the potentially enormous role it may play in the universe as a whole.

Since I suggested that dark matter may be a field, not particles, I want to differentiate it from dark energy.

Though it may only exist as fields, dark energy is spread evenly throughout all of space. It is also known to work against gravity. It pushes things apart, while gravity pulls them together.

Dark matter is not spread evenly throughout space. it exists as clumps, like ordinary matter does, which would indicate that there should be particles. But dark matter only seems to express itself as gravity, not in any other way which particles commonly do.

Dark matter not only does not bump into any known particles, it doesn't affect them through any of the descriptors (or dimensions) we just talked about.

My suggestion is that it may have a concentration like normal particles do, but just doesn't reach into space to form particles, and apparently doesn't reach into the other dimensions either.

Chapter 7 - Conclusion

It's my opinion that the properties of all of our most fundamental particles and fields should be considered to be dimensions, since I believe they fit the definition. What this does for us is to create a model of the universe in which all of physics can emerge from a fundamental substance in the movement dimension interacting with all of the dimensions which make up our world.

This includes our 3 spacial and single time dimension and also dimensions of electric charge, spin, color, and so on. All of these dimensions are quantized as we can tell by their minimum quantities.

Distance and time are in minimum measures of Plancks. Electric charge is in minimum measure of 1/3 of an electron charge. Spin is in 1/2 revolutions. And so on.

The movement dimension cannot be quantized in order to assure us of consistency in our physics. But what we refer to as quantum fluctuations are likely a result of quantization. Our laws of physics may work perfectly in the movement dimension and just show inconsistences in our world due to the grid.

While particles have one real location, their real probability cloud still extends beyond it, allowing for influences not apparent when simply considering a particle.

And, what appear to be multiple paths of particles' probabilities, which ultimately interfere with each other are actually multiple paths of their real clouds which can actually follow multiple paths to create self-interference.

Fields may just be clouds which have no peak entering space to form a particle in our world. But, they may interact with other dimensions such as mass/energy, charge, spin, color, and so on.

In the model I've presented, I tried to incorporate the most widely accepted concepts we have today and fit them together in a way which most likely represents the world we experience. My intent has not been to create new theories but to assemble those we already have with the hope that their assembly leads to new ideas about how things work in our world.

If this model does describe reality, then I think there may not be fundamental particles as we had been searching for. Particles in this model are representations of their fundamental substance. And, the substance of all things is the ether. The study of particles, then, is about why certain ones occur and not others. Why not a whole mix of particles of every size and property? How do they transform into each other? What are the rules of the dimensions?

And, if the ether is the underlying substance, then the most basic laws of physics must be regarding the ether's relationship with our quantized dimensions, the grid. I've attempted to initiate this process, but much work is yet to be done to explain why these particles we have formed and not all other particles which we could imagine. We have a lot of known particles and their behaviors to account for. Perhaps this model can aid in that search.

Regarding the search for fundamentals

In the search for the fundamental pieces of our world and the rules they follow, I think we should expect to find things we're not familiar with, objects not like anything we know.

Having taken objects apart in our exploration of nature, we've always found them to be made of other parts which are not like what they form. Assembled pieces have always formed a new and unique product.

A piece of metal has little resemblance to a car. A molecule has little resemblance to metal. With each layer, we learn of new and unique components. And, we find that those components, somehow, explain the thing they make.

As we look for the most elementary parts of nature, we should expect to find something which is not like anything else, but somehow explains what we do have.

That fundamental component we should expect to remain unexplained. Having no layer beneath it, we have no way to find its explanation. But, even when we're convinced that we've found the most basic substance of our world and the rules it follows, we should still never stop searching for an explanation for it.

About the Author

Timothy Michaels is an artist and writer. Being a technically minded, "cerebral" thinker, he enjoys creating realistic art as well as exploring the world around him through physics. This led him to an explanation of color relationships using physics, which resulted in the development of his Color Calculator. Though he has no formal education in physics, Mr. Michaels has an exceptional gift for logic and an insight into the workings of the universe.

His scientific ideas rely on established theories with evidence, and are explained in a simple manner for general readership. Mr. Michaels provides new ways of understanding these theories, and his ideas go beyond current science to solve problems and make new predictions. In doing so they provide the concepts and formulas necessary to advance science. Students of physics, astronomy, cosmology and other fields will find his explanations enlightening and his new ideas worth investigating.

Other Books by Timothy Michaels

"Absolute Relativity: How Newton and Einstein Agree"

"Out of This World: The Movement Dimension"

"The Physics of Color Harmony"

"How We See Art"

"Walking the Cards: A Unique Drawing Method"

His artwork may be viewed at: www.tmsartgallery.com

Readers are invited to comment by sending an email to: 101timsplace@gmail.com

Bibliography

Bryson, Bill. *A Short History of Nearly Everything.* Broadway Books, 2003

Carroll, Sean. *The Big Picture.* Dutton, 2016

Carter, Rita. *Mapping the Mind,* University of California Press, 1998

Clegg, Brian. *The Universe Inside You,* Icon Books Ltd., 2012

Feynman, Richard. *The Pleasure of Finding Things Out.* Perseus Publishing, 1999

Fong, Peter. *Elementary Quantum Mechanics.* World Scientific Publishing Co. Pte. Ltd., 2005

Frenkel, Edward. *Love and Math,* Amazon Books

Goswami, Amit., Reed, Richard E., Goswami, Maggie. *The Self-Aware Universe,* Penguin Putnam, Inc., 1993

Greene, Brian. *Until the End of Time: Mind, Matter, and Our Search for Meaning in an Evolving Universe,* Alfred A. Knopf Books, 2020

Gunther, Leon. *The Physics of Music and Color.* Springer Science and Business Media, LLC, 2012

Hoffman, Paul. *The Man Who Loved Only Numbers,* Hyperion, 1998

Kaku, Michio. *Hyperspace.* Doubleday, 1994

Kaku, Michio. *The Future of the Mind.* Anchor Boors, 2014

Levenson, Thomas. *The Hunt for Vulcan,* Random House, 2015

Mesler, James, and H. James Cleave II. *A Brief History of Creation,* W. W. Norton & Company, Inc., 2016

Nassau, Kurt. *The Physics and Chemistry of Color*, John Wiley & Sons, Inc., 1983

Norden, Jeanette. *Understanding the Brain*, The Great Courses, 2007

Penrose, Roger. *The Road to Reality.* Vintage Books, a division of Random House, Inc., 2004

Sagan, Carl. *Cosmos.* Ballantine Books, 1980

The World Book Encyclopedia. World Book Inc., 2020

Wolfson, Richard. *Simply Einstein: Relativity Demystified.* W. W. Norton & Company, Inc., 2003